THE FRAGILE RED and orange petals of the hibiscus (*aloalo - Hibiscus rosa-sinensis*), Hawai'i's state flower, last for no longer than two days before they start to wilt and darken. Favoring full sun, the hibiscus juts itself out from its shrubs and looks as if, once picked, it could last forever. The *aloalo* grows most abundantly during the summer months and constant pruning assures a harvest year round.

Opposite page: The Red/Chinese hibiscus (*aloalo - Hibiscus rosa-sinensis*) is the most popular in the hibiscus family and can be found growing nearly everywhere in Hawai'i. The wild and unwieldy trees can grow up to twenty feet if not cut back. The flowers which may be orange, yellow, magenta and pink, are not only beautiful to look at, but are also used in teas, dyes, and for medicinal purposes.

When the hibiscus became the state flower in 1923, it was not the popular full-bodied flower seen today (pictured top) but the rare and exquisite native red hibiscus (*kokio 'ula'ula - Hibiscus kokio*) (middle) found in the remote edges of West Maui.

The fragrant white blossoms of the native white hibiscus (*koki'o ke'oke'o - Hibiscus arnottianus*) (bottom) are scarce and found primarily in private gardens. When in bloom the trees are stunning. Their plump clusters of bright white flowers can be seen from far away, a blazing beacon against the green mountain ranges and blue skies.

THE POINTED ORANGE and purple bracts of the Bird of Paradise (*manu o palekaiko - Strelitzia reginae*) rise out of the sturdy green foliage in the shape of a hungry bird's mouth. A study in graceful and haughty lines, the Bird of Paradise is a picture of floral confidence. Brought to the islands from South Africa, the flower is popular for decorating and can be found in exotic arrangements that can last, if cared for properly, for over two weeks.

These two Bird of Paradise (*manu o palekaiko - Strelitzia reginae*) seem to be speaking as they bend towards each other in a conspiratorial whisper. The brilliant orange flowers blossom up to six during a two week period, a luscious burst of sunshine against the muted green of its hull-like covering. The Bird of Paradise evolved in Madagascar. Its seeds seldomly appear since the flowers are made to be pollinated by birds (not found in Hawai'i). In fact, it is the only flower to be pollinated by birds' feet!

Opposite page: THE THICK POD-LIKE bracts of the giant red heliconia (*Heliconia caribaea*) measure up to five inches long and fall into each other, one on top of the other, in symmetrical perfection.

SPLOTCHES OF PINK rise up over the yellowish-green bracts of the rainbow heliconia (*Heliconia wagneriana*) like a blush. The bracts curve out and end in a pointed and delicate tip. A native of Costa Rica and Panama, the rainbow heliconia thrives in Hawai'i's hot and humid climate. Although the plant needs nothing but lots of sun, water and good soil, it blooms only for a short time. Researchers are now at work on lengthening the March-to-May blooming season.

A FIELD OF flourishing purple vanda orchids (*'okika - Vanda teres XV*) are small and vibrant dots of color against the majestic West Maui mountains. Vanda orchids are a major agricultural export and are sold by the pound for leis and cut flower arrangements. Native from India to Indonesia, vandas range in color from lavender, to purple, to orange.

Opposite page: VIBRANT PURPLE AND rich yellow dendrobium orchids (Dendrobium species) bloom and cascade over their thick and fibrous stalks. These orchids can grow nearly anywhere, from a cold and snowy environment to the warm and humid tropics. Found lining many resort landscapes, as well as in cut flower arrangements in homes, offices and hotels, the cultivation of dendrobiums for the flower trade makes it an important crop.

THE SEDUCTIVE SHAPE of the cattleya orchids (pictured top and middle) signify weddings and other festive occasions. Made up of three outer petals, the flower crinkles and frills, framing the center "lip." The red speckled purple petals of the vanda orchid (*'okika - Vanda teres XV*) are a common sight around the islands and flourish particularly well in the humid and often-times cloud-covered environment on the Big Island. Developed in Singapore in the 1890's, the vanda orchids became well known after Hawai'i growers started mass marketing the flower to the mainland after World War II.

THE THICK, MOSS pocked trunk of the red jade vine (*Mucuna bennettii*) tree twists and turns out of the earth like a rope. Hanging from its branches, brilliant red clusters of bird-like beaks, the red jade vine epitomizes the sultry feel of the tropics. Originally from New Guinea, the red jade now thrives in Hawai'i and can grow to over four feet in gardens big enough to accommodate this lusty vine.

The simple sweetness and vivid yellows and reds of the plumeria flower (*pua melia - Plumeria rubra*) are an integral part of island life. There are seven species and numerous hybrids of this flower. The first plumeria plant ("common yellow") was brought to Hawai'i in 1860. Trees sprout up with ease and anyone can grow one. Just break off a branch, (be careful of the milky sap) allow the end to heal for a couple of days and then plant it deep into a fertilized hole.

Opposite page: HOVERED CLOSE LIKE two hearts, the dewy red anthurium (*Anthurium andraeanum*) flowers are a big part of Hawai'i's tropical glamour. Ranging in size from tiny to large, these flowers are thought to have been brought to the islands in 1889 by English missionary Samuel Damon. The anthurium's kindred—taro, caladium and philodendron— are all from the arum family and share similar characteristics. The heart shaped flower, a "petal" (spathe), surrounds the nubby "phallus" (spadix).

THE TOUGH AND shiny leaves of the dark green and red obake anthurium (*Anthurium sherzerenum*) beg a closer look and touch to make sure that the beautiful flowers are not made out of plastic. Obake, meaning ghost or change in Japanese, is a fitting name for the varied and mysterious color of these flowers. The green part of the flower is what the original flower looked like but, after years of work, horticulturists turned the monochromatic flower into a veritable rainbow. Ranging from white, to red, to green, the colossal leaves on the obake anthuriums can measure up to 12 inches in diameter.

DROPS OF WATER collect on the cool and sleek surface of the pink anthurium (*Anthurium andraeanum*). Anthuriums grow best in the humid and wet climate of hothouses and ninety-five percent of the flowers come from over 200 farms on the windward side of the Big Island. These flowers travel easily and can last after several weeks of being cut.

THE BLOOD ORANGE cups of the African Tulip Tree (*'apelika kulehe - Spathodea campanulata*) lean back from the olive green claws of the blossoming flowers. The large frilly flowers bloom year round, but especially from January to June. More than just beautiful, the African tulip is used for dyes and the hard seedpods are used as material in flower arrangements. Its seeds, paper-thin and willowy silver, are sewn together into a feathery white lei.

THE MAJESTIC KING Protea (*Protea cynaroides*) bursts out of the green, a regal sunburst magnificently captured. The furry center bracts of the King Protea collapse inward while the brilliant red bracts jut out from the sides. Officially the national flower of South Africa, the King Protea has been delighting many flower enthusiasts in Hawai'i. The most glorious of the flowers measure twelve inches across, but typically, the flowers measure from six to seven inches in diameter. The King Protea is most common during the winter months.

THE DIVINE BEAUTY of the cultivated pink water lily (*nymphaea rubra*) rises out of the water, a symbol of spiritual perfection. Native to India, these water lilies are found in ponds across the state, especially prevalent on the island of Kaua'i. Some have an ethereal fragrance, others only open up at night. Thriving in murky, often brown and dirty water, the water lily grows, an unblemished and beautiful object, "rising above," as the Hindus noticed, the mire beneath it.

BALANCED DELICATELY ON the large green leaves below it, the rich yellow *kahili* ginger (*'awapuhi-kahili - Hedychium gardnerianum*) stands a proud and splendid flower. The long red stamens protruding from the blossoms are perches for pollination. While walking through forests across the islands, the aroma from the *kahili* ginger wafts through the air, a sweet and ghostly perfume. Hailing from high atop the Himalayas, the *kahili* ginger was first collected in Hawai'i in 1940 and was named for its similarity to the Hawaiian *kahili,* or royal standard. The red and yellow *kahili* were made from native bird feathers and placed atop tall and glistening poles.